Moonlight

by Grace Hansen

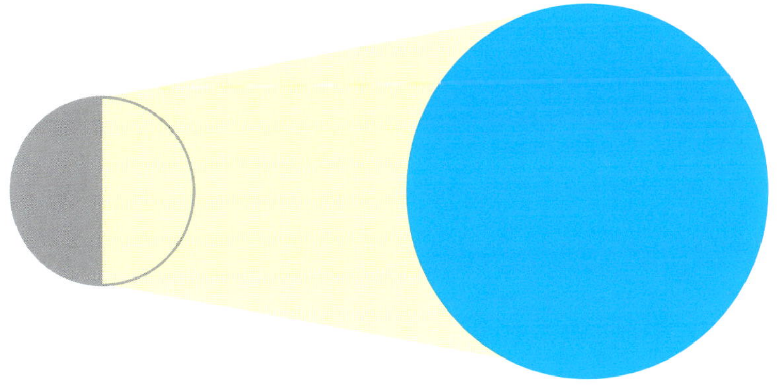

Abdo Kids Jumbo is an Imprint of Abdo Kids
abdobooks.com

abdobooks.com

Published by Abdo Kids, a division of ABDO, P.O. Box 398166, Minneapolis, Minnesota 55439.

Abdo Kids Jumbo™ is a trademark and logo of Abdo Kids.

Printed in the United States of America, North Mankato, Minnesota.

102019

012020

Photo Credits: Alamy, iStock, NASA, Shutterstock

Production Contributors: Teddy Borth, Jennie Forsberg, Grace Hansen
Design Contributors: Dorothy Toth, Pakou Moua

Library of Congress Control Number: 2019941262

Publisher's Cataloging-in-Publication Data

Names: Hansen, Grace, author.

Title: Moonlight / by Grace Hansen

Description: Minneapolis, Minnesota : Abdo Kids, 2020 | Series: Sky lights | Includes online resources and index.

Identifiers: ISBN 9781532189081 (lib. bdg.) | ISBN 9781532189579 (ebook) | ISBN 9781098200558 (Read-to-Me ebook)

Subjects: LCSH: Moon--Brightness--Juvenile literature. | Moon--Phases--Juvenile literature. | Space--Juvenile literature. | Astronomy--Juvenile literature. | Light--Juvenile literature.

Classification: DDC 523.32--dc23

Table of Contents

What Is Moonlight?

The moon shines in the night sky. But the moon does not create its own light. It **reflects** the sun's light!

Moon Phases

The moon looks different throughout the month. These differences are called phases.

Phases are made by the moon's **orbit** around Earth. The moon's orbit takes about 29.5 days to complete.

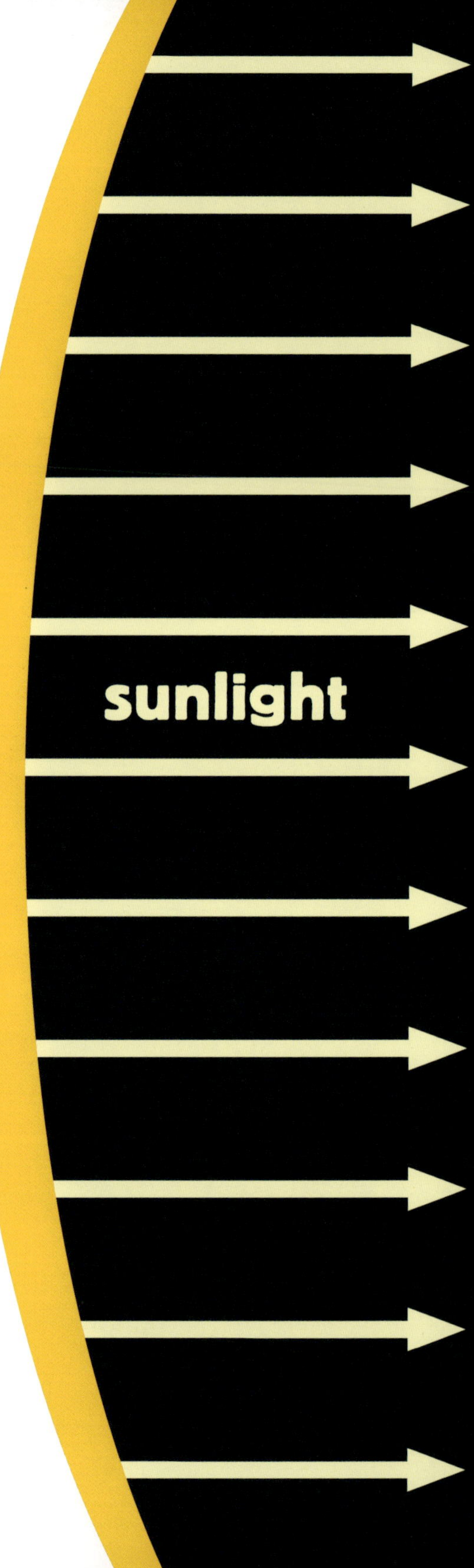

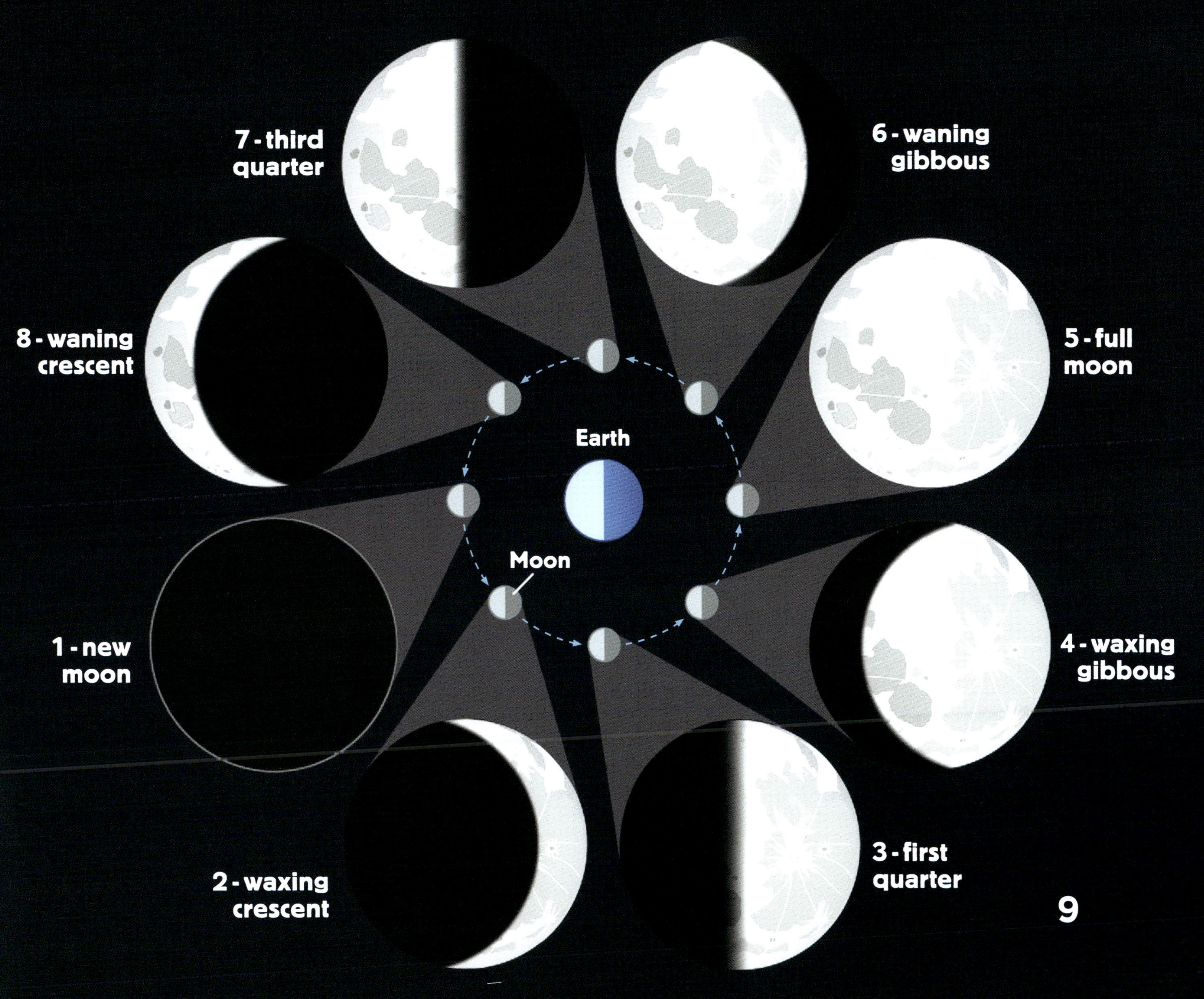
7 - third quarter
6 - waning gibbous
8 - waning crescent
5 - full moon
Earth
Moon
1 - new moon
4 - waxing gibbous
2 - waxing crescent
3 - first quarter

Early in the month, the moon sits between the earth and the sun. The sun's light hits the back of the moon. The side of the moon pointed toward Earth is dark. This is called a new moon.

In a new moon, there is no moonlight. But that changes as the days go by.

In the coming days, the moon's light increases. This is called a waxing moon.

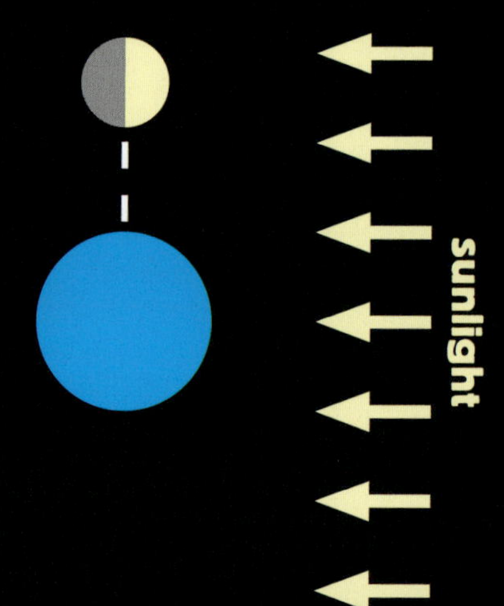

By mid-month, the moon is halfway through its **orbit** of Earth. Now, the earth is between the moon and the sun. The entire sunlit part of the moon is facing us.

A fully lit moon is called a full moon! It is beautiful to look at. But it has some dark spots. These areas are giant **craters** on the moon's surface.

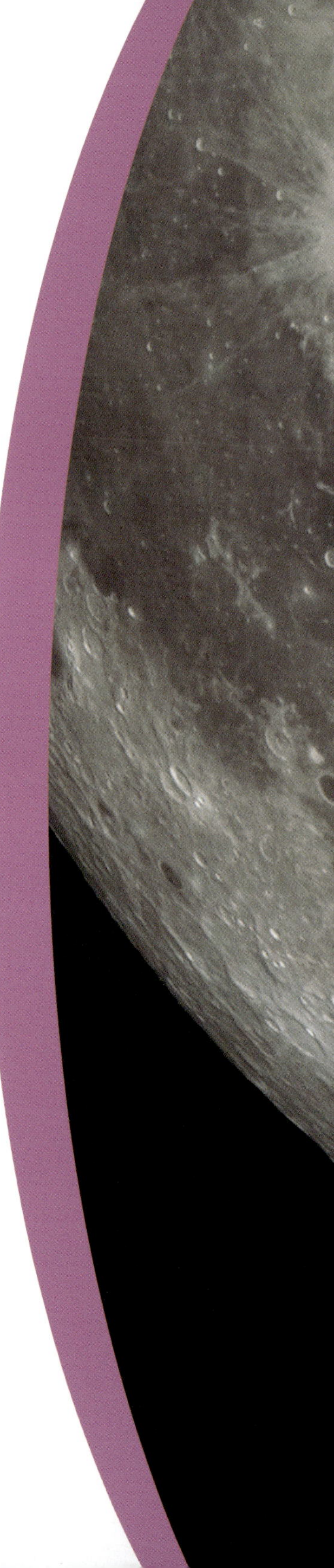

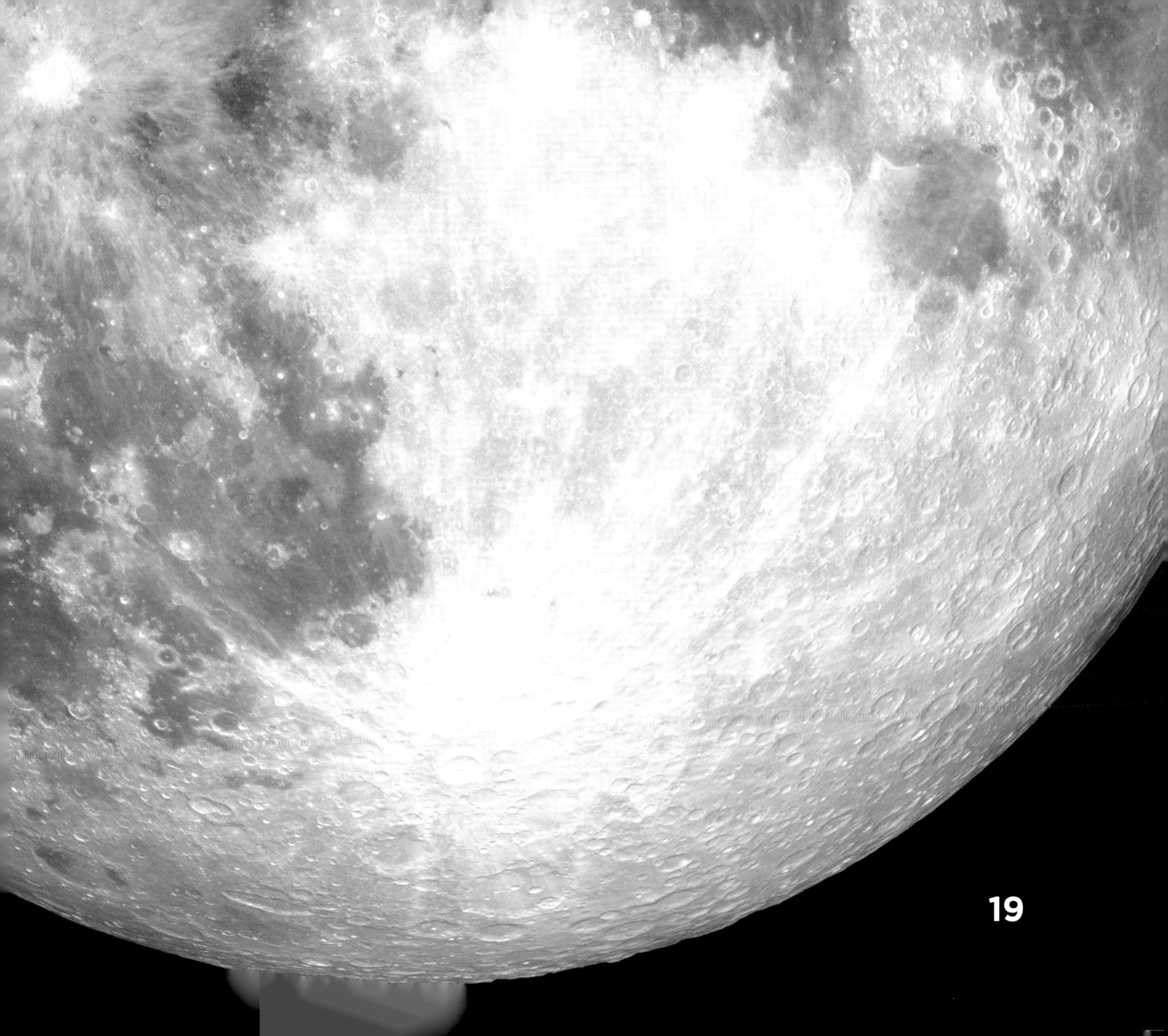

Toward the end of the month, the moon's light is decreasing. This is called a waning moon. It will be a new moon again soon. And we can watch it shine more each night!

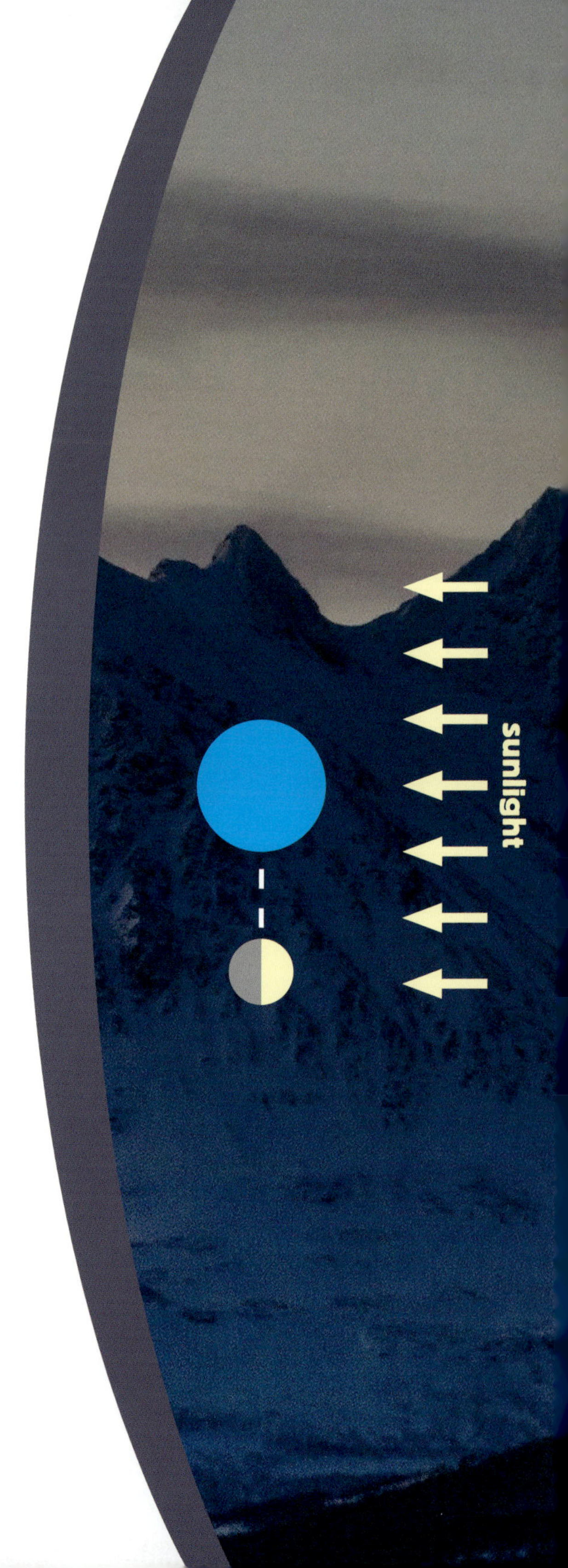

More Facts

- The moon's **craters** have been around for a very long time. This is because the moon does not have weather. Things are **preserved** much longer there than on Earth.

- There are 8 official moon phases. They are new moon, waxing crescent, first quarter, waxing gibbous, full moon, waning gibbous, third quarter, and waning crescent.

- A lunar eclipse can happen when the earth is between the sun and moon. This is when Earth's shadow covers the moon.

Glossary

crater – a hallow area shaped like the inside of a bowl. The moon has many craters on its surface from large impacts with other space bodies, like asteroids.

orbit – a curved path in which a planet or other space body moves in a circle around another body.

preserved – maintained in its original state or existing state.

reflect – a body throwing back light without absorbing it.

Index